This book belongs to
_____

NUMBER SEQUENCE

Write in the numbers that come between the first and last number in each row.

1									10

3									12

5									14

7									16

2									11

4									13

9									18

8									17

6									15

SKIP COUNTING BY 5

Write the numbers counting by 5.

5				
				50
55				
				100
				125

NUMBER ORDER

Write in the numbers in numerical order.

2	3	4	1				
5	3	7	2				
6	2	4	3				
9	6	8	7				
4	6	5	1				
8	7	6	1				

BEFORE AND AFTER

Write the numbers that come before and after.

__	2	__	__	13	__
__	4	__	__	15	__
__	6	__	__	10	__
__	8	__	__	11	__
__	7	__	__	14	__
__	5	__	__	19	__

ADDING ONE MORE

Find the sum.

0 + 1 =

1 + 1 =

2 + 1 =

3 + 1 =

4 + 1 =

5 + 1 =

6 + 1 =

7 + 1 =

8 + 1 =

9 + 1 =

ADDING TWO MORE

Find the sum.

0 + 2 =

1 + 2 =

2 + 2 =

3 + 2 =

4 + 2 =

5 + 2 =

6 + 2 =

7 + 2 =

8 + 2 =

9 + 2 =

ADDING DOUBLES

Find the sum.

0 + 0 =

1 + 1 =

2 + 2 =

3 + 3 =

4 + 4 =

5 + 7 =

6 + 7 =

7 + 7 =

8 + 8 =

9 + 9 =

ADDITION UP TO 10

Find the sum.

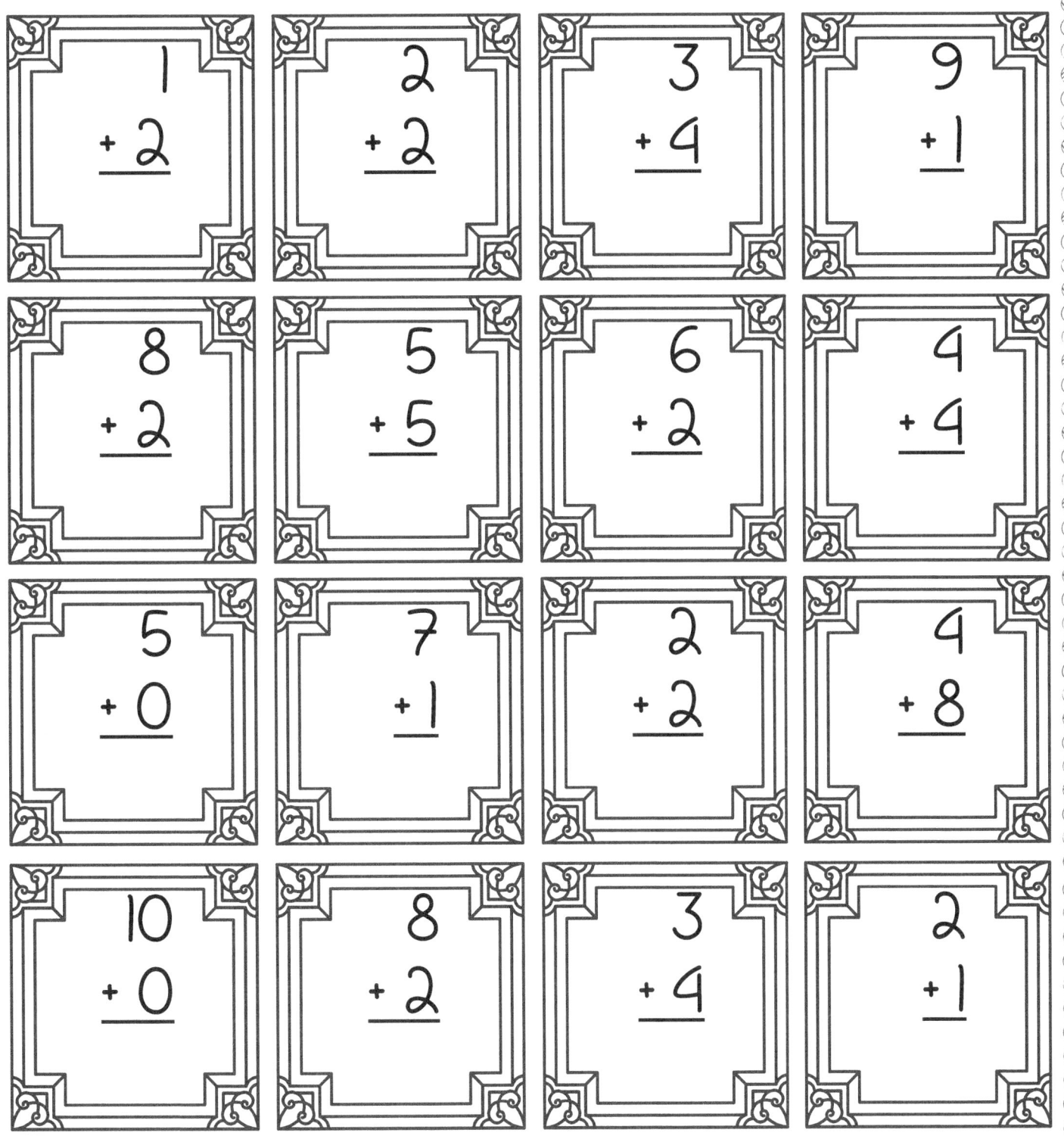

ADDITION UP TO 20

Find the sum.

10 + 2	12 + 2	13 + 4	9 + 9
18 + 12	15 + 4	16 + 4	8 + 10
15 + 2	17 + 1	9 + 5	8 + 7
10 + 6	13 + 2	18 + 1	10 + 10

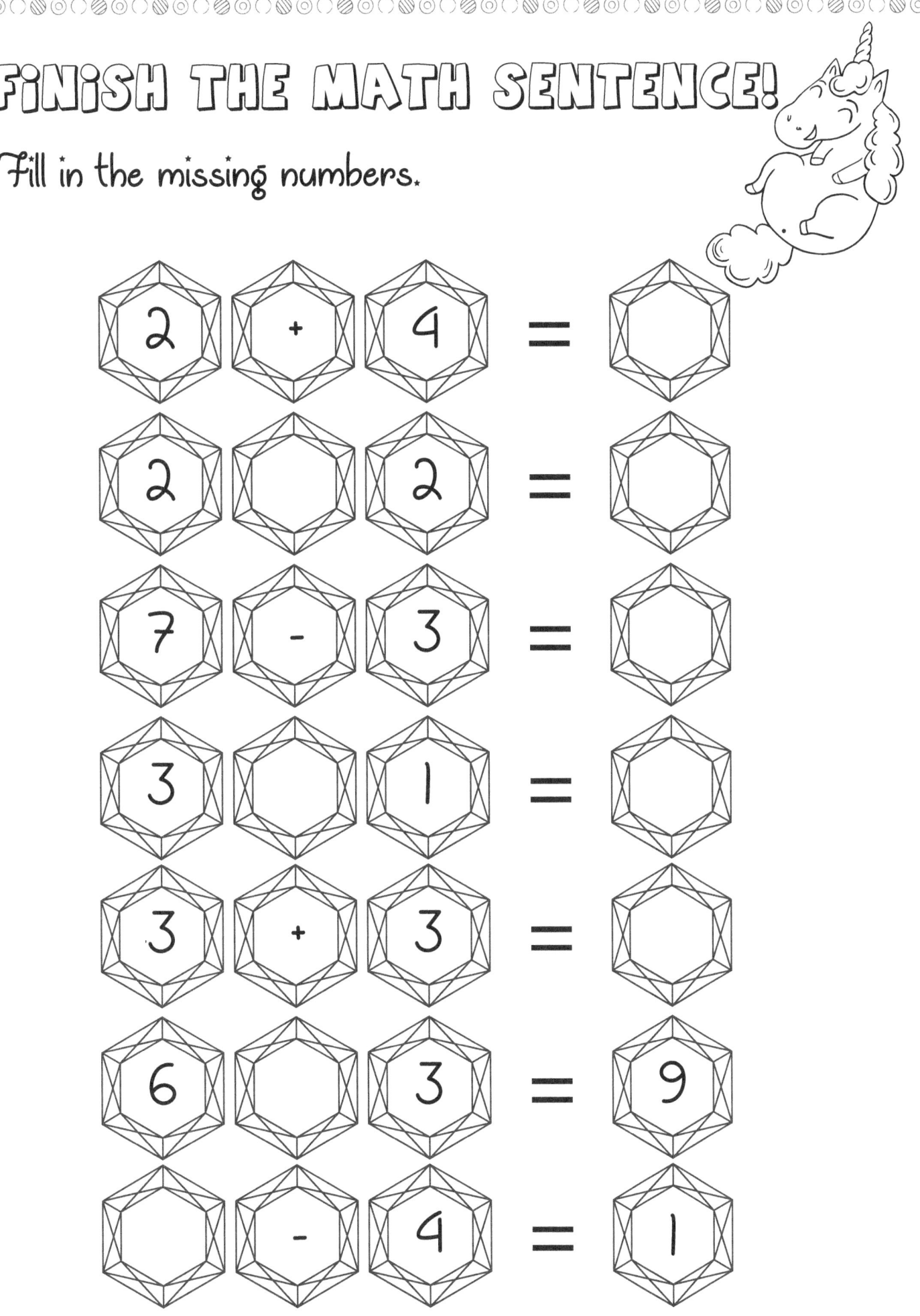

ODD OR EVEN?

Color the numbers using the chart.

ODD - BLUE
EVEN - RED

3 1 12 31

10 13 5 14

7 8 17 22

15 2 25 11

6 4 34 18

NUMBER ORDER

Write in the order each object comes in.

| first | second | third | fourth | fifth |
| sixth | seventh | eighth | ninth | tenth |

MISSING PIECES

Solve the addition problems by filling in the missing numbers.

4 + ☐ = 8	5 + ☐ = 8	6 + ☐ = 8	7 + ☐ = 8
1 + ☐ = 9	2 + ☐ = 9	3 + ☐ = 9	4 + ☐ = 9
3 + ☐ = 10	4 + ☐ = 10	5 + ☐ = 10	6 + ☐ = 10

TEN MORE, TEN LESS

Write in 10 more and 10 less than each number in the middle

	15			60	
	20			22	
	33			85	
	50			75	
	44			38	

TENS AND ONES

Look at the number of the beginning of each row. Write the correct numbers of tens and ones in the boxes next to it.

25 = [2] tens + [5] ones

12 = [ ] tens + [ ] ones

31 = [ ] tens + [ ] ones

19 = [ ] tens + [ ] ones

27 = [ ] tens + [ ] ones

45 = [ ] tens + [ ] ones

50 = [ ] tens + [ ] ones

COUNTING TALLY MARKS

Write in how many Count and write how many tally marks.

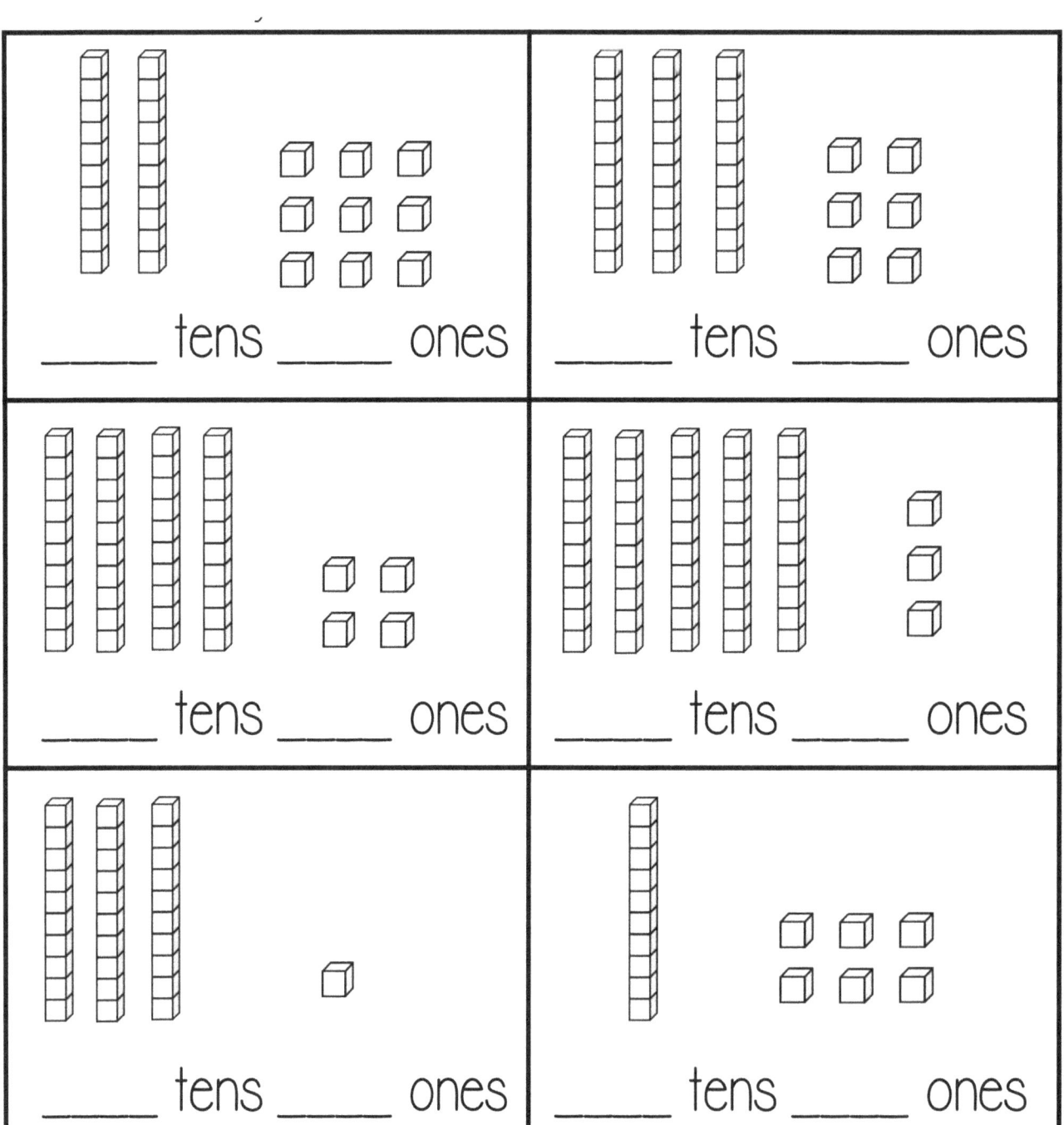

PLACE VALUE MATCH

Draw a line from the blocks to the correct number.

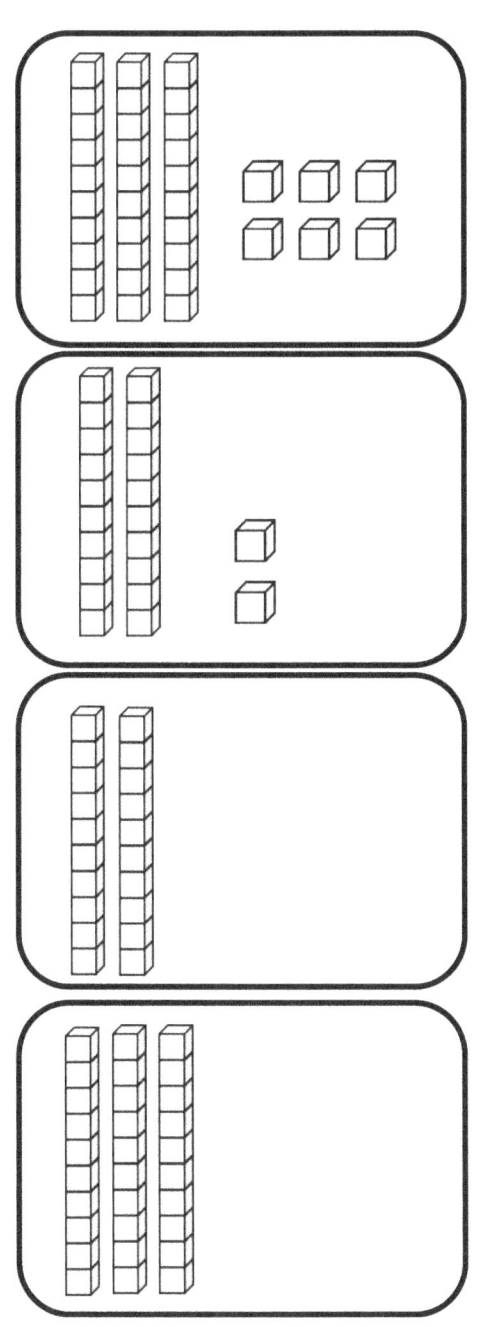

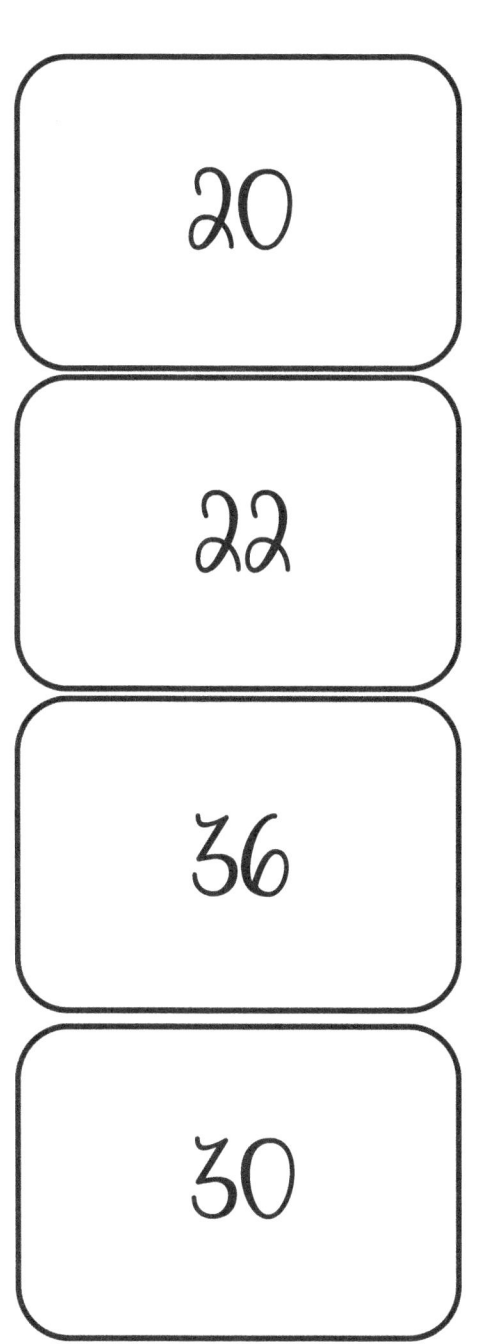

COUNTING TALLY MARKS

Count and write how many tally marks.

|||| ____	|||| |||| ____	|||| ||| ____
|||| ____	|||| || ____	|||| ||||	____

Create a matching amount of tally marks.

15	9	13
6	16	4

BAR GRAPH

Color a place on the graph for the correct amount of each item.

🍦	5	💎	7	🧁	1
🌠	3	♡	4	🍬	2

WRITE THE NUMBER

Instructions: Fill in the missing numbers on the chart.

COUNTING TO 100

Instructions: Trace the numbers from 1 to 100

1	2	3	4	5	6	7	8	9	10
11	12	13	14	15	16	17	18	19	20
21	22	23	24	25	26	27	28	29	30
31	32	33	34	35	36	37	38	39	40
41	42	43	44	45	46	47	48	49	50
51	52	53	54	55	56	57	58	59	60
61	62	63	64	65	66	67	68	69	70
71	72	73	74	75	76	77	78	79	80
81	82	83	84	85	86	87	88	89	90
91	92	93	94	95	96	97	98	99	100

COUNTING TO 100

Instructions: Fill in the missing numbers.

1	2			5	6		8		10
11				15	16		18		20
		23	24		26			29	
31	32					37			40
			44	45					
51			54		56		58		60
61	62				66		68		70
				75	76				80
81	82						88	89	90
91				95					100

MISSING NUMBERS

Cut out and past the missing numbers in each row.

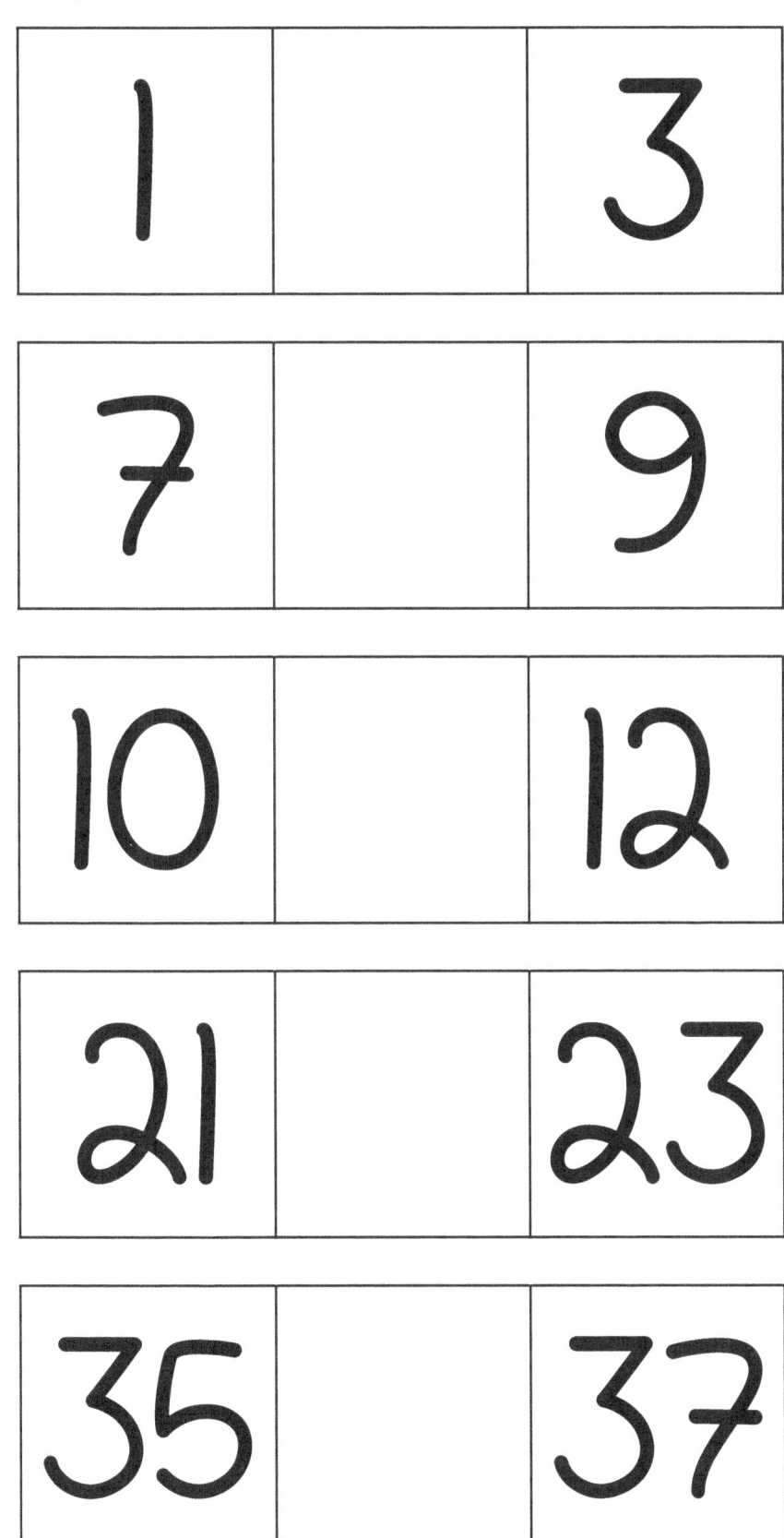

NAME THAT NUMBER

Cut out and paste the number name into the right place.

1	2	3	4	5

6	7	8	9	10

ten	four	two	nine	seven
eight	five	three	one	six

NAME THAT NUMBER

Cut out and paste the number into the right place.

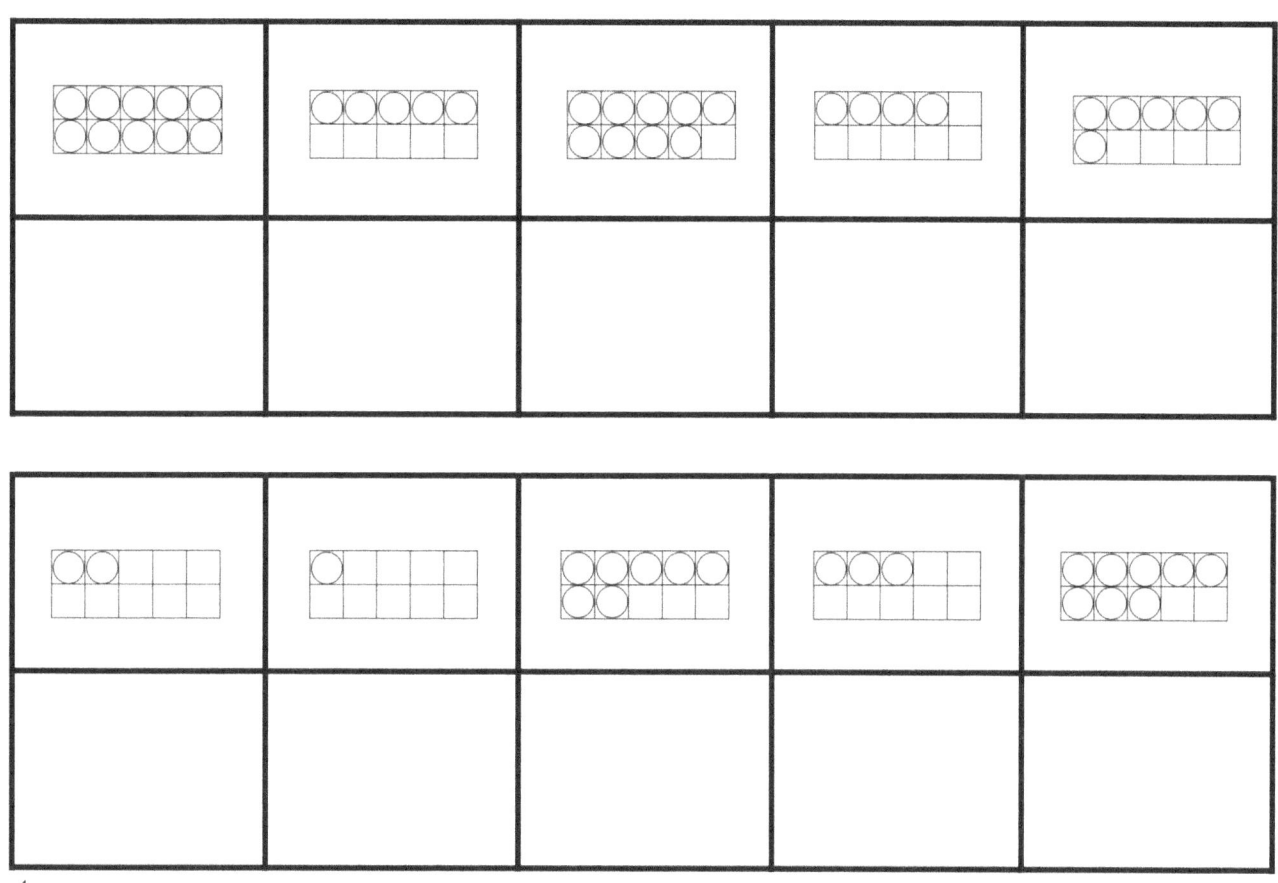

NAME THAT NUMBER

Cut out and paste the number name into the right place.

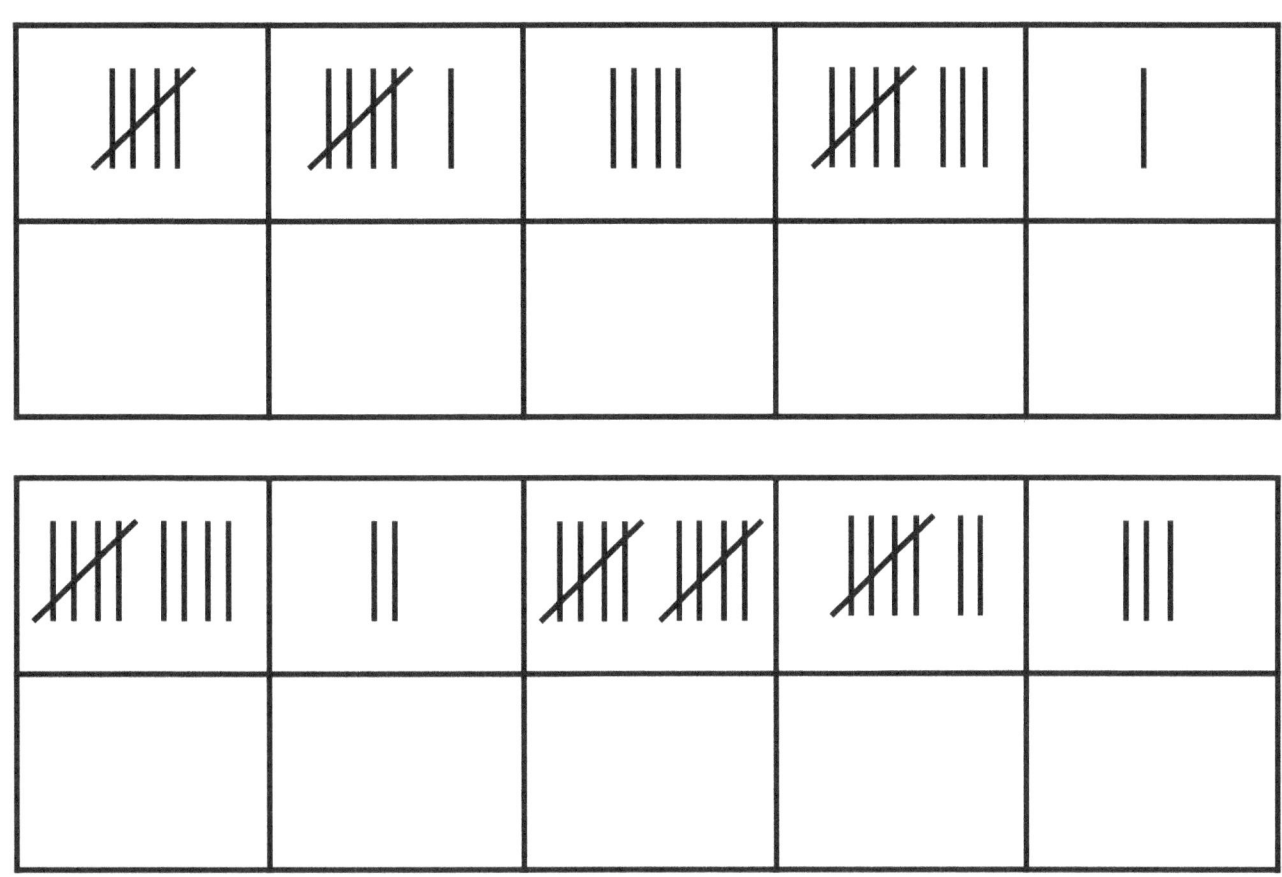

two	seven	nine	three	five
six	four	one	eight	ten

COUNTING FORWARDS

Cut out and paste missing numbers counting forwards.

1		3	4	
	7			10

COUNTING BACKWARDS

Cut out and paste missing numbers counting backwards.

10			7	
5			2	1

2	8	8	6	4
9	6	9	3	5

ODD OR EVEN?

Cut out and paste the numbers in the correct section.

ODD

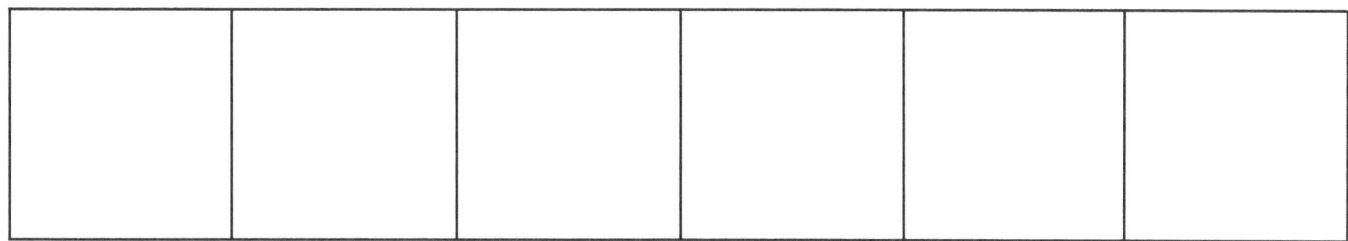

 EVEN

| 7 | 11 | 8 | 9 | 3 | 12 |
| 2 | 1 | 4 | 6 | 5 | 10 |

PLACE VALUE COMPARE

Cut out and paste the more than or less than arrow in the correct box.

Greater than and less than symbols can be used to compare numbers and expressions. The greater than symbol is >. The less than symbol is <.

19 ☐ 25	23 ☐ 25
8 ☐ 2	13 ☐ 33
11 ☐ 17	20 ☐ 27
9 ☐ 4	55 ☐ 51

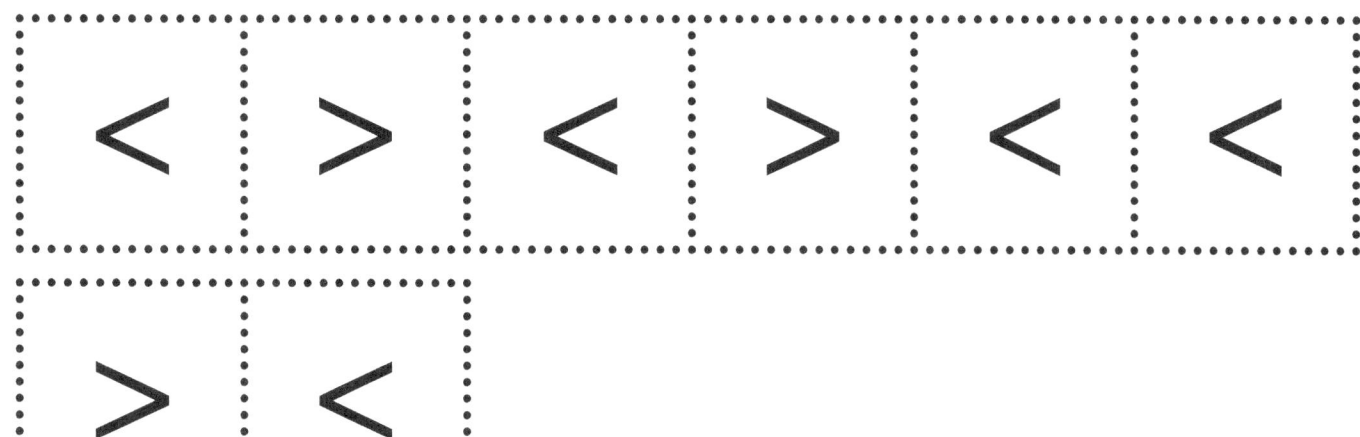

TEN FRAME PUZZLES

Instructions: Cut out the puzzles. Match the number to the correct 10 frame.

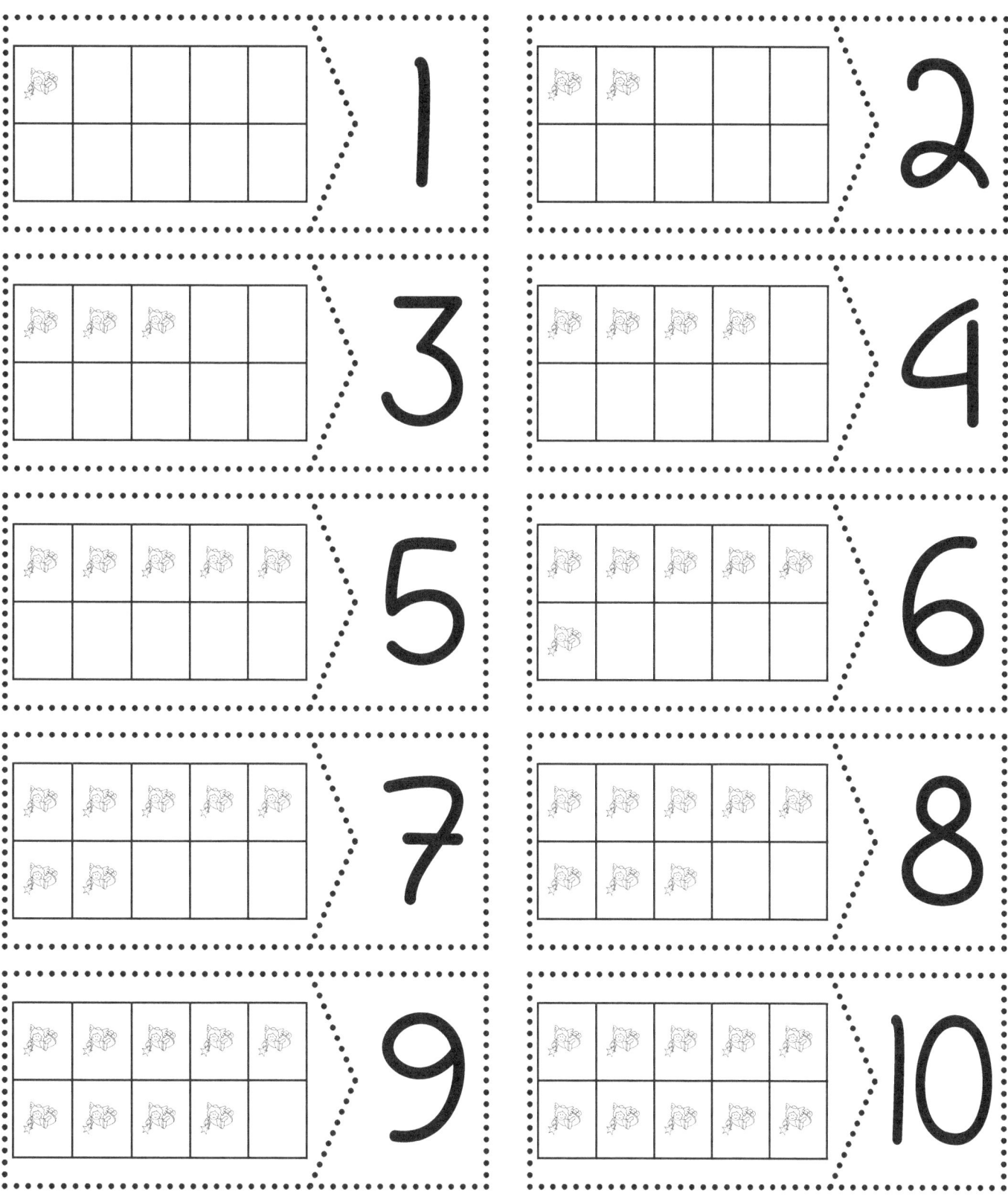

www.ingramcontent.com/pod-product-compliance
Lightning Source LLC
Chambersburg PA
CBHW081647220526

45468CB00009B/2575